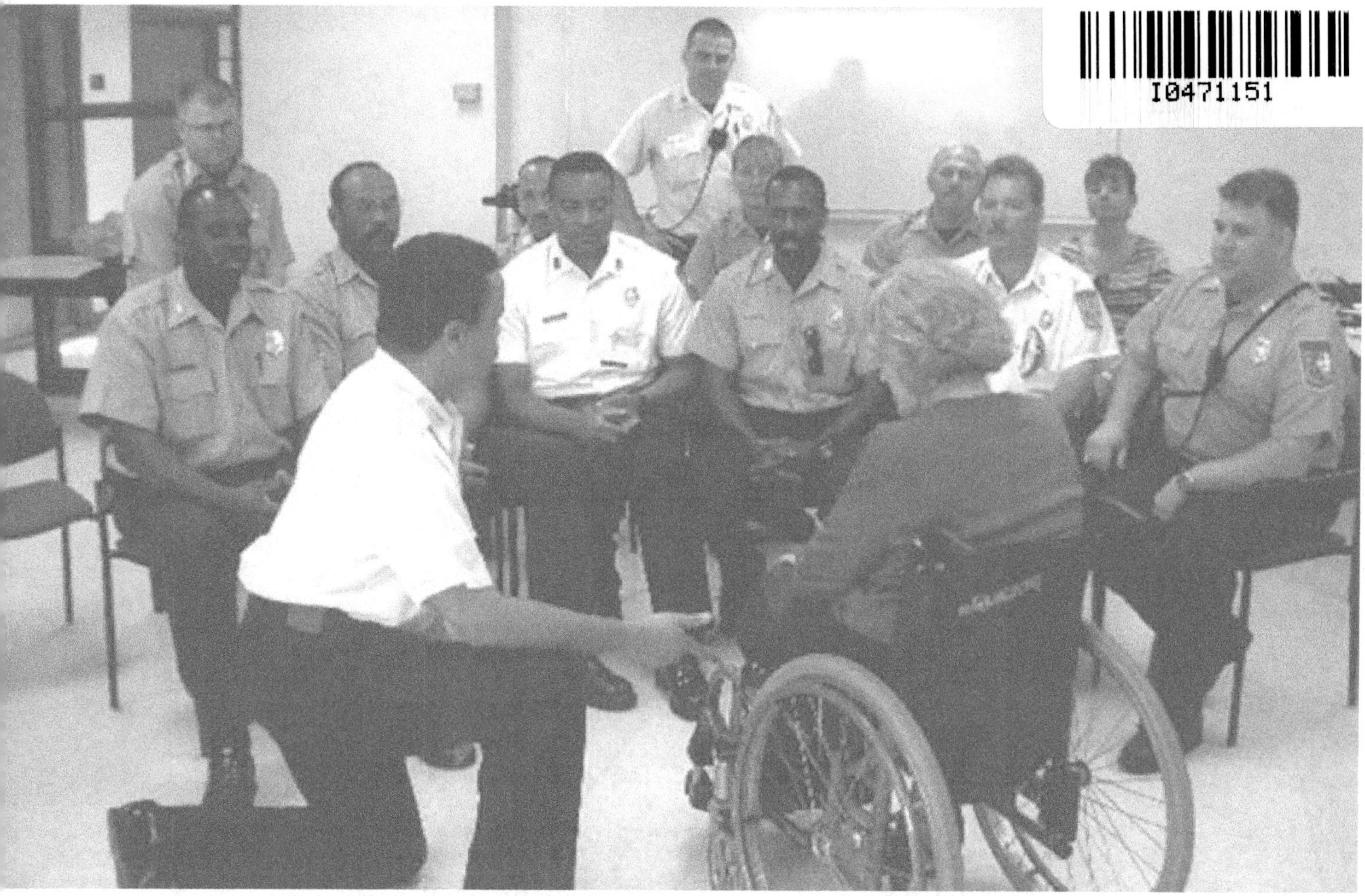

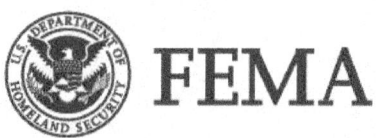

Orientation Manual for First Responders on the Evacuation of People with Disabilities

FA-235 / August 2002

U.S. DEPARTMENT OF HOMELAND SECURITY
FEMA

U.S. Fire Administration
Mission Statement

As an entity of the Federal Emergency Management Agency, the mission of the United States Fire Administration is to reduce life and economic losses due to fire and related emergencies, through leadership, advocacy, coordination, and support. We serve the Nation independently, in coordination with other Federal agencies, and in partnership with fire protection and emergency service communities. With a commitment to excellence, we provide public education, training, technology, and data initiatives.

Contents

Acknowledgements

Letter from the National Organization on
Disability (N.O.D.) 1

Perspectives on Emergency Preparedness 3
   People with Disabilities 3
   First Responders 3
   Identifying Those with Special Needs 5
   Locating People with Disabilities in Your Community
   to Include in Emergency Preparedness Planning 6

Categories of Impairments 7
   Vision Impairments 7
   Dog Guides 8
   Hearing Impairments 9
   Cognitive Impairments 11

Mobility 13
   Wheelchair Users 14
   Ambulant with Aide 14
   Comparison of Carry Techniques 15
   Evacuation Devices 19

Conclusion 21

Appendix A: Writing Guidelines and
Disability Terminology 23

Appendix B: Resources 25

Appendix C: Disaster Mitigation
for Persons with Disabilities 27

Appendix D: TTY – Commonly Used Abbreviations 29

Appendix E: References 31

Acknowledgements

This guide was developed by the United States Fire Administration. Members of the following Review Panel met to determine the flow, content, and organization of the guide and reviewed all sections of the guide for clarity.

Special thanks to members of the Baltimore County Fire Department, Dundalk Station 6 and Eastview Station 15 for their demonstrations and photos. Also, thanks to the Community College of Baltimore County-Dundalk, (formerly the Dundalk Community College) 7200 Sollers Point Rd., Baltimore County, MD, for the use of their facilities during the photo shoot.

For additional copies of this publication, or to obtain a copy in a different format, write to, call, or visit:

United States Fire Administration
16825 South Seton Avenue
Emmitsburg, Maryland 21727
(1-800-561-3356)
www.usfa.fema.gov/USFAPUBS

Review Panel

Edwina Juillet
Fire & Life Safety for People with Disabilities

Brian Black
Eastern Paralyzed Veterans Association

Marianne Cashatt
Disability Awareness/Public Relations

Alan Clive
Federal Emergency Management Agency

Marsha Mazz
U.S. Architectural and Transportation Barriers Compliance Board

Sharon Davis
The Arc of the United States

Captain Elgin Browning
Orange (TX) Fire Department

Captain Glenn Blackwell
Baltimore County (MD) Fire Department

Curtis Robbins, Ph.D.
S & C Robbins Association

Lisa Codario
National Science Foundation

Fall 2002

To All First Responders and Other Emergency Personnel:

The National Organization on Disability (N.O.D.) applauds the United States Fire Administration and the Federal Emergency Management Agency for their joint commitment to publish and distribute this *Orientation Manual for First Responders on the Evacuation of People with Disabilities*.

While individuals with disabilities have always been aware of their own unique needs during emergency situations, the September 11th, 2001 terrorist attacks have inspired a new commitment to partnering with the emergency response community. This led N.O.D. to establish the *Emergency Preparedness Initiative* to encourage these communities to jointly work toward effective techniques that will save lives - - all lives - - when moments counts.

This *Orientation Manual* provides practical information: both rescue techniques, and preparedness information for the first responder who will encounter people with visible and non-visible disabilities through their work. This publication can assist first responders in advance or make critical emergency decisions. In conjunction with its companion piece *FA-154 Emergency Procedures for Employees with Disabilities in Office Occupancies*, this publication will help first responders to confidently work with the disability community toward the goal of protecting and saving lives and minimizing trauma.

We hope first responders will incorporate this information into their arsenal of skills and experience to benefit all people they serve. With awareness about the unique issues that may impact a specific segment of the population during an emergency, a first responder's ability to react appropriately and identify creative solutions increases exponentially to everyone's benefit.

N.O.D. salutes all emergency personal and especially the first responders for your generous and continued commitment of service to the public.

Sincerely,

Alan A. Reich
President
National Organization on Disability

Elizabeth A. Davis
Director
Emergency Preparedness Initiative

Perspectives on Emergency Preparedness

People with Disabilities

Three months after terrorist attacks redefined American life, most of the country's 54 million citizens with disabilities said that they did not feel sufficiently prepared for future crises. According to Harris Interactive survey results released by The National Organization on Disability (N.O.D.):

- 58 percent of people with disabilities say they do not know who to contact about emergency plans for their community in the event of a terrorist attack or other crisis.

- 61 percent say that they have not made plans to evacuate their homes quickly and safely.

Among those who are employed full- or part-time, 50 percent say that no plans have been made to evacuate their workplace safely.

"The country as a whole has some catching up to do to be prepared, but these statistics show that people with disabilities lag behind everyone else. This is a critical discrepancy, because those of us with disabilities must in fact be better prepared so we are not at a disadvantage in an emergency," said N.O.D. President Alan A. Reich.

"It is critical that all plans for emergency preparedness consider the special needs of people with disabilities. We strongly advocate that individuals with disabilities themselves be included in the planning process at all levels," said Reich.

First Responders

It is incumbent on first responders to learn how best to perform a rescue using equipment and procedures that facilitate safe evacuation for any person with a disability.

People with disabilities are entitled to the same level of protection in an emergency as everyone else—no more, no less. The "reasonable accommodation" mandated in the Americans with Disabilities Act (ADA) is intended to provide the same level of safety and utility for people with disabilities as is provided to everyone. Key points to consider are as follows:

- Remember that a person with a disability has unique abilities and limitations. Accommodations should be made that emphasize their abilities.
- Include the person in the decision-making process when selecting special equipment and design of procedures.
- Familiarize yourself with the building and occupants in your response area; identify coworkers, neighbors, and friends who can be of assistance in an emergency.
- Participate in the development of emergency evacuation planning for the occupancies in your community before the emergencies occur.

People with disabilities are increasingly moving into the mainstream of society, which contributes to the diversity that has been this country's strength. Further, we cannot predict when anyone of us may need assistance, such as in the case of a broken leg or the development of heart or lung disease.

Does your Local Emergency Planning Committee (LEPC) include individuals with disabilities as active members?

Do your Emergency Operation Plans (EOP) take into consideration individuals' capability to evacuate with little or no assistance?

Have members of your fire or emergency services department had awareness training in evacuating people with disabilities?

Which of these techniques is *not* appropriate for individuals with disorders of the back, spine, or spinal cord?

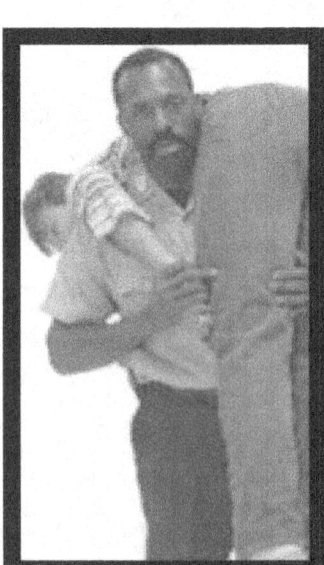

Identifying Those with Special Needs

First responders need to take into consideration that there are many individuals who are protective of their right to independence and privacy and who may be reluctant to be identified.

In the development of your plans, think of each individual as one who happens to have a particular disability. Do not make the mistake of lumping together all persons with disabilities into one group or category.

A Lesson Learned

Prior to the February 1993 World Trade Center bombing, people with disabilities said that in the interest of privacy or because they felt that they did not need special assistance, they had opted not to be identified as disabled in the emergency management plan.

After the bombing, they realized that they did need special assistance; they had not realized how vulnerable they were outside of normal working hours when there were few coworkers available.

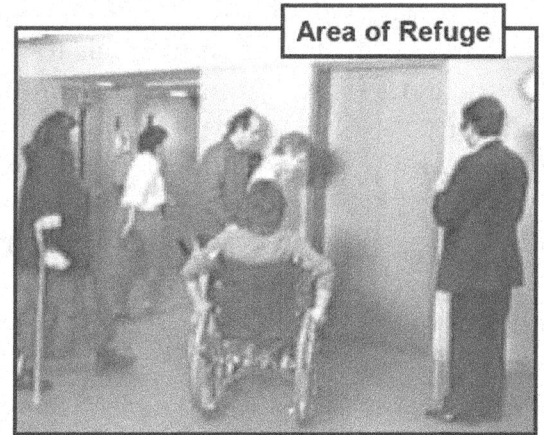
Area of Refuge

For example, there are building emergency evacuation plans and codes on which they were based, containing instructions for all persons with disabilities to go to an area of refuge and wait for members of the emergency team to escort them to safety.

As a general rule, there is no reason that a person who is blind or deaf cannot use the stairs to make an independent escape as long as he or she can effectively be notified of the need to evacuate and can find the stairway.

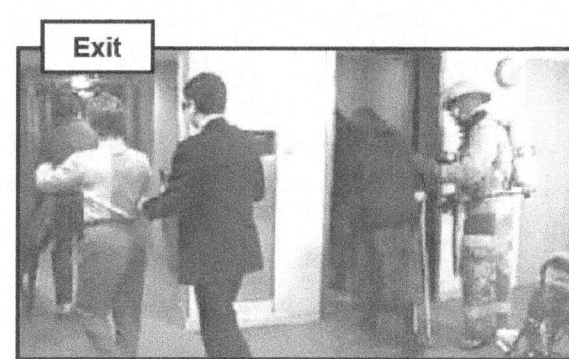
Exit

Locating People with Disabilities in Your Community to Include in Emergency Preparedness Planning

1. Contact national disability organizations and/or their local affiliates. (See Appendix B, Resources.)

2. Contact your State Vocational and Rehabilitation Agency. Part of their work is to introduce volunteer and public service organizations to their clients.

   a) View the Federal Consumer Information Center's online list of State Vocational and Rehabilitation Agencies.

   b) Search the Internet using a phrase such as "[your state] Vocational Rehabilitation."

   c) Look in the Government pages of your local phone book for state or local listings such as "Rehabilitation Services Administration" or "Rehabilitation Information."

3. There are several hundred Centers for Independent Living (CIL) across the country. CILs are community-based resource and advocacy centers managed by and for people with disabilities, promoting independent living and equal access for all persons with physical, mental, cognitive, and sensory disabilities. (See illustration below.)

4. Contact your state or local government's committee, commission, or council on disabilities. These are often part of the Governor's office or cabinet.

5. Contact Department of Veterans Affairs facilities in your state, which serve people with disabilities.

6. Contact your local ADA coordinator, who usually can be located through the Mayor's office or county government office.

7. Contact local church congregations, which may be aware of specific community members with disabilities.

8. Ask professionals who serve people with disabilities—such as special education teachers, or occupational, physical, or speech therapists—if they can suggest individuals to participate in emergency planning. You might try contacting the National Rehabilitation Association.

— N.O.D., February 20, 2002

To find the CIL serving your locale, go to www.virtualcil.net/cils, and click on a state.

Categories of Impairments

Vision Impairments
(Blind or Low Vision)

When you approach a person who has vision impairments, announce your presence. Speak naturally and directly to the individual not through a third party. Say something like, "Hi, I am Firefighter Jones." Describe the action to be taken, then ask the person to tell you the best way to assist him or her.

First responders should take steps to ensure that after exiting the building, individuals with impaired vision are not abandoned but are led to a place of safety. Someone should remain with them until the emergency is over.

A lesson learned from the World Trade Center 1993 bombing involved the complaints of blind tenants who, after being escorted down and out of the building, were unceremoniously left in unfamiliar environs in the midst of a winter ice storm, where they had to negotiate ice-covered sidewalks and falling glass from overhead.

Tips for Assisting People with Visual Impairments

- It's okay to use words like "see," "look," or "blind." Let the individual grasp your arm or shoulder lightly for guidance. He or she may choose to walk slightly behind you to gauge your body reactions to obstacles.

- Be sure to mention stairs, doorways, narrow passages, and ramps.

- When guiding to a seat, place the person's hand on the back of the chair.

- If leading several individuals at the same time, ask them to hold each other's arms.

Dog Guides

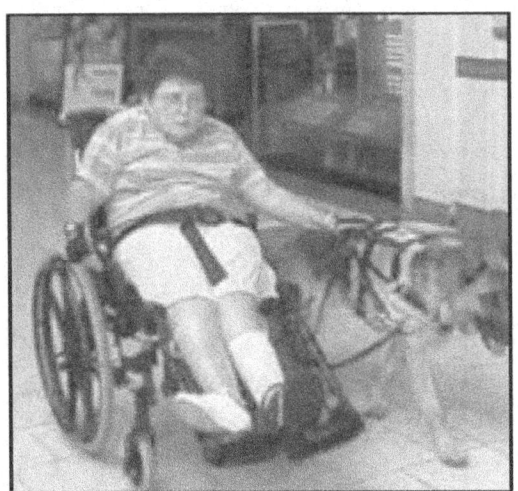

Wheelchair Dog

Traditionally, the term "service animal" referred to seeing-eye dogs©★ for people with vision impairments. However, today there are many other types of "service animals." In addition to guide dogs for the blind, there are:

- Hearing dogs for people who are deaf

- Seizure dogs for people who have seizure disorders

- Assist animals for people with motor impairments

- Companion animals for people with psychiatric impairments

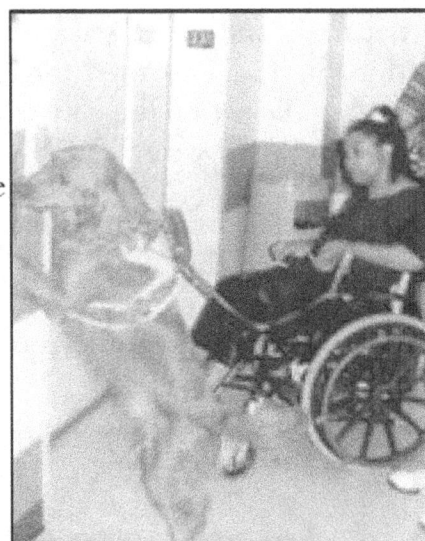

Quad Dog

★The term, "Seeing-eye dog©," is a copyrighted term used by the school of that name. Dog guides for the blind are just that, dog guides. There are corporate names such as "seeing eye," "leader dog," and "guide dog."

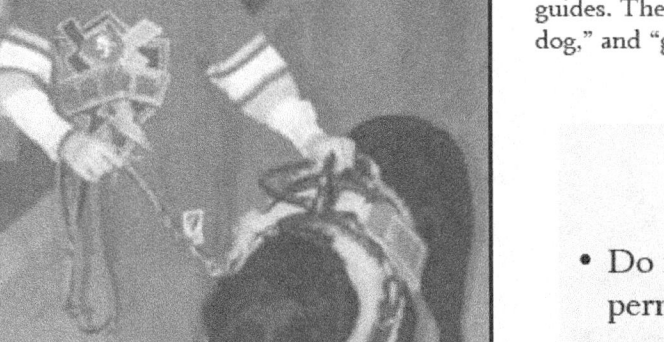

Walker Dog

Tips for Assisting Owners with Service Dogs

- Do not pet or offer the dog food without the permission of the owner.

- When the dog is wearing its harness, he is on duty. In the event you are asked to take the dog while assisting the individual, it is recommended that you hold the leash and not the dog's harness.

- Plan for the dog to be evacuated with the owner.

Hearing Impairments
(Deaf or Hard of Hearing)

Hearing impairments range from mild hearing loss to an extreme or profound deafness, the level at which an individual receives no benefit from aural input. Many persons who are hearing impaired can use their residual hearing effectively with assistance from hearing aids or other sound amplification devices, often augmented by lip reading.

However, hearing aids also amplify background sounds. For example, the sound of emergency alarms interferes with or even drowns out voice announcements, rendering the emergency voice communication system useless. An accommodation for this problem is to provide individuals with a tactile/vibratory pager, which is tied into the building fire notification system. When the audible alarm is activated, they are simultaneously notified by the vibration of the pager.

Another major problem of emergency notification faced by persons who are deaf or hard of hearing is that they cannot keep up with storm warnings on radio and television. In Florida, TV stations now close-caption their hurricane coverage, and several deaf organizations have set up TDD/TTY hotlines to keep persons up-to-date on storm warnings and evacuation orders.

When approaching a person who is deaf or hard of hearing, and the person does not have visual contact with you, e.g., if you are entering a room, flick the lights on and off to draw attention to your presence.

If you are rescuing a person who is deaf or hard of hearing, do not allow others to interrupt or joke with you while conveying the emergency information. Be patient; the individual may have difficulty comprehending the urgency of your message.

Tips for Assisting People with Hearing Impairments

- Face the light. Do not cover or turn your face away, and never chew gum.

- Establish eye contact with the person, even if an interpreter is present.

- Use facial expressions and hand gestures as visual cues.

- Check to see if you have been understood, and repeat, if necessary.

- Offer pencil and paper. Write slowly, and let the individual read as you write. Written communication may be especially important if you are unable to understand the individual's speech.

- Provide the individual with a flashlight for signaling his or her location, in the event that he or she is separated from the rescuing team or buddy, and to facilitate lip-reading in the dark.

WPSD Students' Book Helps EMS Workers

By David Brodbeck, Staff Writer

From the newsroom of the Woodland Hills Progress Star,
Woodland Hills, Pennsylvania, Wednesday, April 17, 2002 .....

An emergency scene is chaotic enough, but throw in not being able to communicate with people and the situation becomes problematic.

To aid emergency medical personnel in their dealings with people who are unable to hear, students attending Western Pennsylvania School for the Deaf in Edgewood have put together a book of helpful tips.

Gretchen Steffen, residential supervisor at WPSD, started the project two years ago as part of a community service project with Baldwin Borough's EMS.

To help the EMS department, the students have put together a book that has photographs of them demonstrating sign language that may be encountered when responding to an emergency.

The Baldwin paramedics keep a copy of the book in each of their ambulances and each paramedic has a copy to study, Steffen says.

The Pleasant Hills Lions Club has expressed interest in working to get the guide developed by the WPSD students circulated throughout the county.

http://www.baldwinems.com/

A training class was held to give four of the students an opportunity to interact with paramedics. By working together, the students and paramedics learned from each other. Afterwards, the students and paramedics requested to schedule follow-up classes to develop stronger communication skills.

Cognitive Impairments

Tips for Assisting People with Cognitive Impairments

- Be patient.
- Break down information into simple steps.
- Use simple signals or symbols.
- Do not talk about a person to others in front of him or her.
- Do not talk down to or treat the person as a child.

- Provide pictures, symbols, or diagrams instead of words.
- Read written information.

- Provide written information on audiotape.
- Use voice output on the computer.
- Use Reading Pen single words (get illustration of Reading Pen).
- Use line guide to identify or highlight one line of text at a time.

People with developmental disabilities may experience limitation with cognitive abilities, motor abilities, and social abilities. The individual may have difficulty in recognizing rescuers or being motivated to act in an emergency.

These individuals may also have difficulty in responding to instructions which involve more than a small number of simple actions. Keep in mind that:

- Visual perception of written instructions or signs may be confused.

- Sense of direction may be limited, requiring someone to accompany them.

- Ability to understand speech is often more developed than his or her own vocabulary.

- The individual should be treated as an adult who happens to have a cognitive or learning disability.

Fire.

Come with me.

Go down the stairs.

*Definition of developmental disability: severe, chronic disability attributable to mental or physical impairment, which becomes apparent before the person is 22 years old, and is likely to continue indefinitely. The disability can result in substantial limitation in three or more areas: self-care, receptive and expressive language, learning, mobility, self-direction, capacity for independent living and economic self-sufficiency, as well as the continuous need for individually planned and coordinated services.

Use of Pictures to Replace Spoken Language

Below is a sample of a product of pictographs to replace the spoken language.

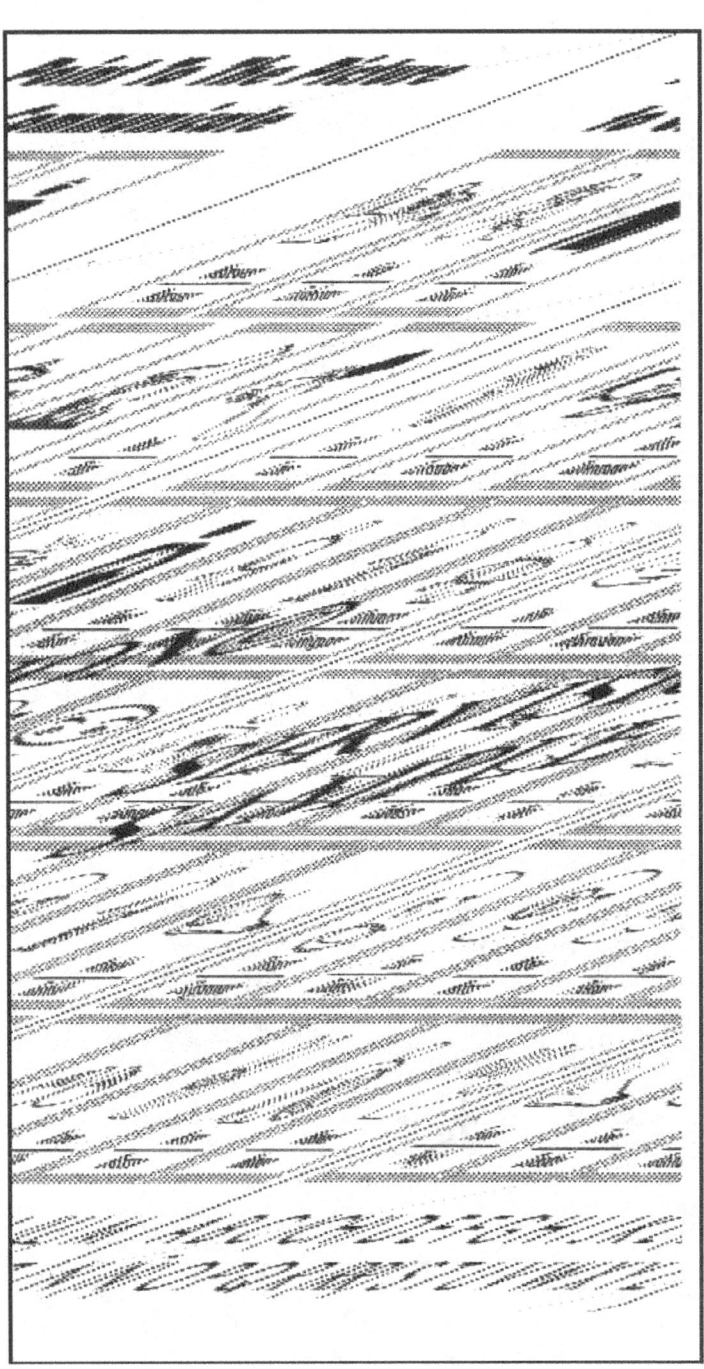

Care-Receiver

This communication board is to be used to assist you in expressing your needs in times of disaster or other emergency situations. If you are the victim of such circumstances and are having problems expressing your needs, simply point to the picture or phrase that represents your situation or identify an item you need.

Care-Giver

This communication board has been designed to bridge communication gaps in times of disaster or other emergency situations. The gaps may result from language barriers, disability, age, or the trauma and confusion associated with the event. As a result, critical information could be difficult to exchange. By either acknowledging or pointing to the appropriate picture, you can better assess health conditions.

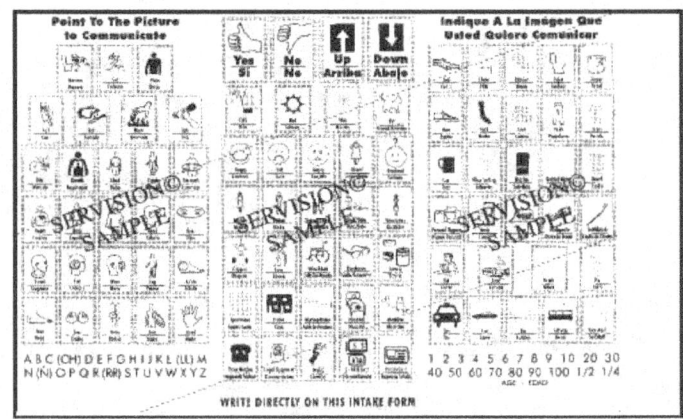

Samples courtesy Servison, Inc.,
982 N.E. 126th Street, North Miami, FL 33161, Phone: 305-899-1996.

Mobility

Wheelchair Users

Because people who use wheelchairs have a wide variety of abilities and limitations, it is difficult to generalize their needs.

Following are some useful questions to consider in understanding common limitations.

- Is the person able to stand or walk without the aid of the wheelchair?

- If yes, how long can the person stand or walk unaided?

- Does the person have full, partial, or no use of the upper extremities?

Wheelchair users are trained in special techniques to transfer from one chair to another. Depending on their upper body strength, they may be able to do much of the work themselves. As with persons with vision impairments, ask first what you can do to help them.

During rescues in which the person is transferred from his or her wheelchair into an evacuation chair, the first responder needs to consider that once the person has been taken to a place of safety, the wheelchair should be waiting for him or her, if possible. People who depend on their wheelchairs for mobility have expressed concern about their wheelchairs being left behind.

If you assist a wheelchair user, avoid putting pressure on the person's extremities and chest. Such pressure might cause spasms, pain, and even restrict breathing. For example, carrying someone slung over your shoulders (like the old and now incorrect "fireman's carry") is like sitting on a person's chest and poses grave danger for individuals who fall within categories of neurologic and orthopedic disabilities.

Ambulant with Aide

Tips for Assisting Wheelchair Users

- When giving directions to a person in a wheelchair, consider distance, weather conditions, and physical obstacles such as stairs, curbs, and steep hills.

- Relax and speak naturally. Do not be embarrassed if you happen to use accepted common expressions such as "got to be running along" that seem to relate to the person's disability.

- When addressing a person who uses a wheelchair, do not lean on the wheelchair unless you have permission to do so. A wheelchair is part of an individual's personal space.

- When talking to a person who uses a wheelchair, look at and speak directly to that person, rather than through a companion.

- When talking with a person in a wheelchair for more than a few minutes, use a chair whenever possible. This can facilitate conversation.

- Terms such as "wheelchair bound" or "confined to a wheelchair" are inappropriate. Using a wheelchair does not mean confinement.

- When greeting a person who uses a wheelchair, it is appropriate to offer to shake hands with that person even if he or she has upper extremity limitations.

Someone using a walking aide, such as a crutch or a cane, might be able to negotiate stairs independently. One hand is used to grasp the handrail, while the other hand is used for the crutch or cane. Here, it is best not to interfere with this person's movement. You might be of assistance by offering to carry the extra crutch. Also, if the stairs are crowded, you can act as a buffer and run interference.

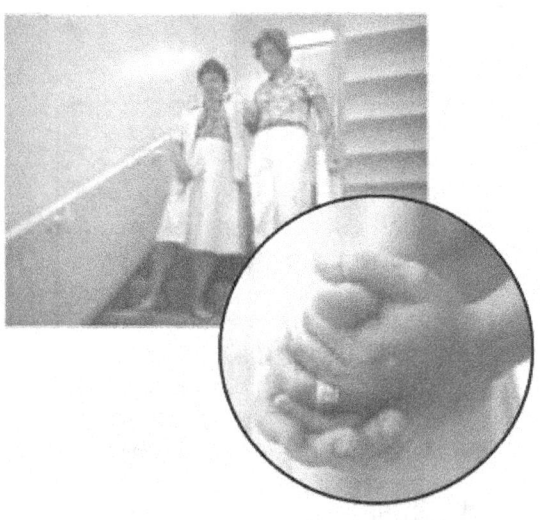

Comparison of Carry Techniques

Cradle Carry

The Cradle Carry method should be used when the person has little or no arm strength. It is safest if the person being carried weighs less than the carrier's weight.

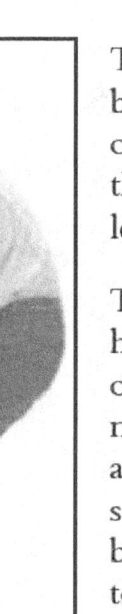

The choice of a particular carry has little to do with the disability of the person to be rescued, and more to do with his or her size, abilities, and certain problems, such as leg spasms. The choice between the carries relates then to the general physical attributes of both the rescuer and the rescuee.

As the size of the rescuee approaches the size of the rescuer, the difficulty of choosing a technique increases and becomes more and more dependent on the abilities of the rescuee.

Swing- or Chair-Carry

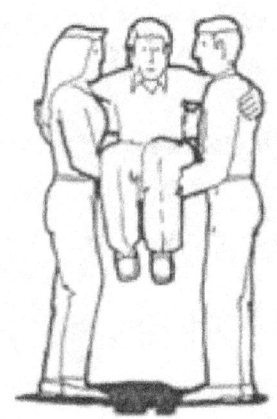

The advantage of the Swing- or Chair-Carry is that partners can support, with practice and coordination, a person whose weight is the same or even greater than their own weight. The disadvantage is increased awkwardness in vertical travel (stair descent) due to the complexity of the two-person carry. Also, three persons abreast may exceed the effective width of the stairway.

Stand on opposite sides of the individual being rescued.

Take the rescuee's arm on your side and place it around your shoulder.

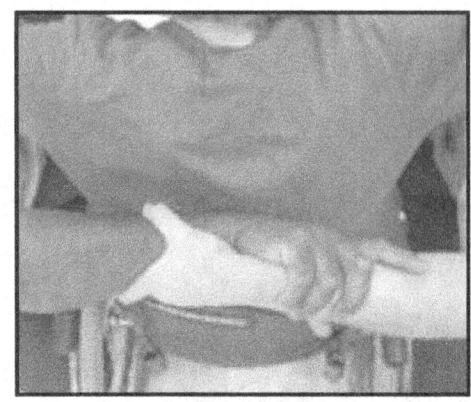

Grasp your carry partner's forearm behind the rescuee, at the small of his or her back.

Reach under the rescuee's knees and grasp your carry partner's other wrist.

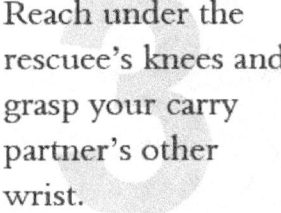

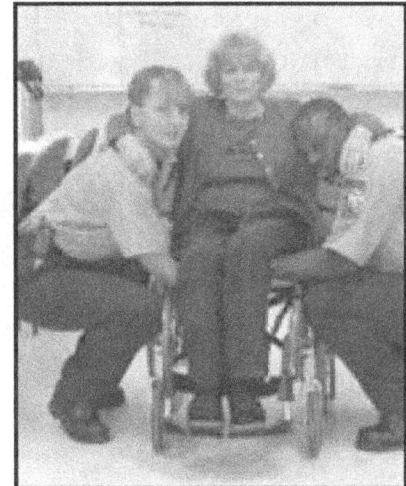

Lean in close to the individual and lift at the count of three.

Continue pressing into the individual being carried to provide additional support.

After completion of the lift, shift the rescuee upward for a more comfortable carry.

Note that the person being carried has her forearms resting on the shoulders of the carry partners.

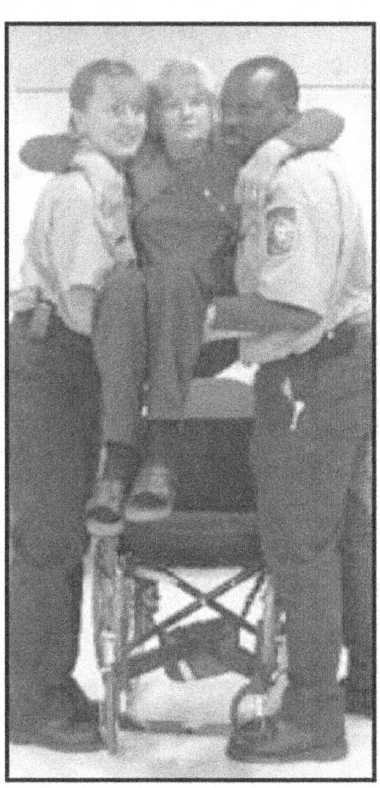

In-Chair Carry

The wheelchair user is anxious to be returned to his or her wheelchair after the rescue; therefore, the in-chair evacuation is desirable when feasible.

One-Person Assist

- Grasp the pushing grips, if the wheelchair has them.
- Stand one step above and behind the wheelchair.
- Tilt the wheelchair backward until a balance (fulcrum) is achieved.
- Keep your center of gravity low.
- Descend frontward.
- Let the back wheels gradually lower to the next step.
- If possible, have another person assist you.

Two-Person Assist

Positioning of second rescuer:

- Stand in front of wheelchair.
- Face the wheelchair.
- Stand one, two, or three steps down (dependent on height of the rescuer).
- Grasp the frame of the wheelchair.
- Push into the wheelchair.
- Descend stairs backward.

The person in front should be careful not to lift the wheelchair, as this places additional weight on the person assisting behind the wheelchair.

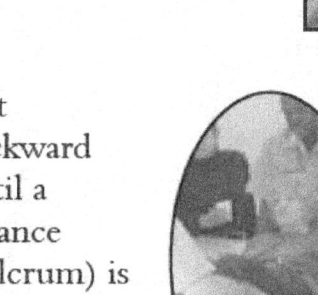

Tilt backward until a balance (fulcrum) is

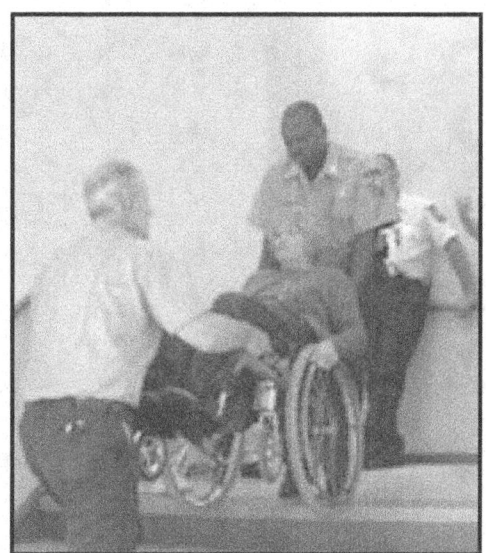

Note: wheelchair user is grasping inside (chrome) wheel control, thereby

Three-Person Assist

Position for second and third rescuers:

- Face direction of descent.
- Flank the wheelchair.
- Stay in line with the two front (smaller) wheels.
- Stand one step/tread lower than rescuer behind wheelchair.
- Grasp the frame of the wheelchair.
- Push into the wheelchair.

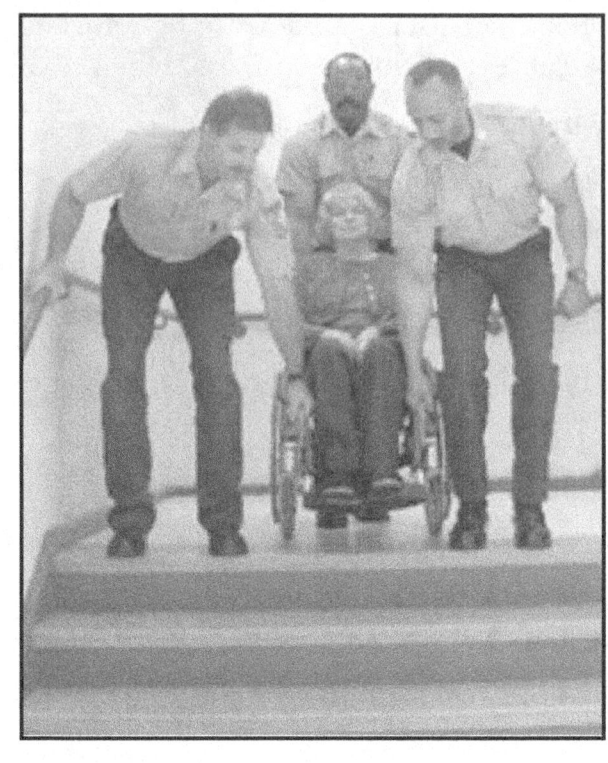

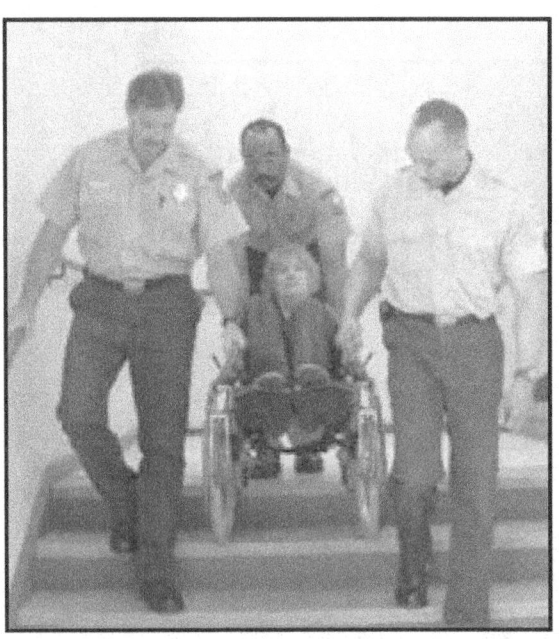

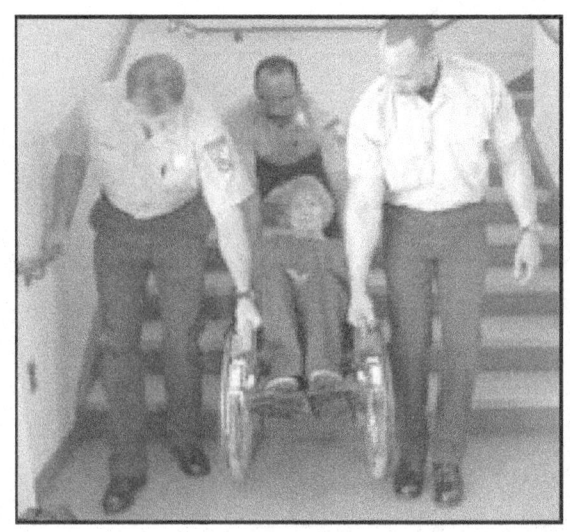

Evacuation Devices

"...it is because of this chair and those that risked their lives to bring me down that I am able now to write these words..." John Abruzzo

When the first plane hit the tower on September 11, 2001, John Abruzzo, like others, rushed to the stairwell. However, evacuation for John would prove to be much more difficult than others, as he is a C 5-6 quadriplegic who relies on an electric wheelchair for mobility. With use of the emergency evacuation chair, John was able to make an escape to safety. "It took us an hour and a half to get down 69 floors."

In the first attack on the World Trade Center in February 1993, John's evacuation took six hours, during which he was carried in his electric chair from the 69th floor to the 44th floor, where he was transferred to a stretcher and taken out of the building. Shortly after the 1993 bombing, the Port Authority purchased a number of products and systems to aid in the evacuation and life-safety of the World Trade Center occupants. The implementation of these products proved successful on 9/11. Lights stayed on while John and his friends evacuated, ventilation systems in the stairwells minimized smoke infiltration, and the evacuation chairs became a real life-saver.

Several other people with disabilities were successfully evacuated with evacuation chairs. John and his group exited the tower and were out of harm's way just ten minutes before its collapse.

Some Surprised to Find Themselves Heroes
By Bruce Horovitz, USA TODAY (September 12, 2001)

On the 68th floor, they came upon a distraught woman in a wheelchair. They did not know her name, but the woman, who was blond and in her 40s, told them she had been in the building during the terrorist bombing there in 1993.

The two men helped her out of (her) wheelchair into a special chair designed for emergency evacuations stored in the stairwell. (The chair had been purchased after 1993 bombing.) They strapped her in and began the journey down as the building blazed. Others pitched in to help carry from time to time. Along the way, they passed dozens of firemen running up the stairway. And they passed a number of aged or overweight people who could not keep up pace. They stumbled outside the building one hour later and placed the woman in an emergency van.

Conclusion

The issues regarding persons with disabilities are many and complex. This guide is simply a first step to give you, the first responder, some insight and guidance to better serve the unique needs of this community of people. The information found in Appendix B can take you to the next level.

To paraphrase the National Organization on Disability printed earlier in the guide, we hope you will incorporate this information into your arsenal of skills and experience to benefit all people that you serve. With awareness about the unique issues that may impact a specific segment of the population during an emergency, your ability to react appropriately and identify creative solutions increases exponentially to everyone's benefit.

On behalf of all of us who contributed to the material, we thank you for taking the time to read this guide, and also for your many contributions to all communities.

Appendix A

Writing Guidelines and Disability Terminology

When writing, it's important to be concise, particularly in journalism. However, sometimes the effort to limit wordiness leads to inappropriate references to people with disabilities. The following guidelines explain preferred terminology and reflect input from over 100 national disability organizations. These guidelines have been reviewed and endorsed by media and disability experts throughout the country. Although opinions may differ on some terms, the guidelines represent the current consensus among disability organizations. Portions of the guidelines have been adopted into the *Associated Press Stylebook*, a basic reference for professional journalists.

Do not focus on disability unless it is crucial to a story. Avoid tear-jerking human interest stories about incurable diseases, congenital impairments, or severe injury. Focus instead on issues that affect the quality of life for those same individuals, such as accessible transportation, housing, affordable health care, employment opportunities, and discrimination.

Do not portray successful people with disabilities as superhuman. Many people with disabilities do not want to be "hero-ized." Like many people without disabilities, they simply wish to live lives of full inclusion in our communities and do not want to be judged based on unreasonable expectations.

Do not sensationalize a disability by writing "afflicted with," "crippled with," "suffers from," "victim of," and so on. Instead, write "person who has multiple sclerosis" or "man who had polio."

Do not use generic labels for disability groups, such as "the retarded," "the deaf." Emphasize people, not labels. Say, "people with mental retardation" or "people who are deaf."

Put people first, not their disability. Say, "woman with arthritis," "children who are deaf," "people with disabilities." This puts the focus on the individual, not the particular functional limitation. Despite editorial pressures to be succinct, it is never acceptable to use "crippled," "deformed," "suffers from," "victim of," "the retarded," "the deaf and dumb," etc.

Emphasize abilities, not limitations. For example:

- **Correct:** "uses a wheelchair," "uses braces," or "walks with crutches."
- **Incorrect:** "confined to a wheelchair," "wheelchair-bound," or "crippled."

Similarly, do not use emotional descriptors such as "unfortunate," "pitiful," and similar phrases.

Disability groups also strongly object to using euphemisms to describe disabilities. Terms such as "handi-capable," "mentally different," "physically inconvenienced," and "physically challenged" are considered condescending. They reinforce the idea that disabilities cannot be dealt with directly and candidly.

Do not imply disease when discussing disabilities that result from a prior disease episode. People who had polio and experienced after-effects have a post-polio disability; they are not currently experiencing the disease. Do not imply disease with people whose disability has resulted from anatomical or physiological damage (e.g., person with spina bifida or cerebral palsy). Reference to disease associated with a disability is acceptable only with chronic diseases, such as arthritis, Parkinson's disease, or multiple sclerosis. People with disabilities should never be referred to as "patients" or "cases" unless the relationship with their doctor(s) is under discussion.

Show people with disabilities as active participants of society. Portraying persons with disabilities interacting with people without disabilities in social and work environments helps break down barriers and open lines of communication.

The Research Training Center/Independent Living (RTC/IL) NIDRR acknowledges the National Institute on Disability and Rehabilitation Research (NIDRR) for providing funds to develop the first edition of these guidelines.

Appendix B

Resources

National Organization on Disability (N.O.D.) *Actions taken as a result of the events of 9/11*	910 Sixteenth Street, N.W., Suite 600 Washington, DC 20006 Voice: 202-293-5960 e-mail: ability@nod.org Web: www.nod.org	National Organization on Disabilities promotes the full and equal participation and contribution of America's 54 million men, women, and children with disabilities in all aspects of life.
The Access Board *Technical assistance in matters of ADA and ADAAG*	1331 F Street, NW, Suite 1000 Washington, DC 20004-1111 Voice: 202-272-0080; 800-872-2253 TTY: 202-272-0082; 800-993-2822 e-mail: Info@access-board.gov Web: www.access-board.gov	An independent federal agency devoted to accessibility for people with disabilities. Click on *Evacuation Planning*. For information about disability groups, go to: www.access-board.gov/links/disability.
ABLEDATA *Assistive technology, emergency evacuation equipment*	8630 Fenton Street, Suite 930 Silver Spring, MD 20910 Voice: 800-227-0216 TTY: 301-608-8912 e-mail: abledata@macroint.com Web: www.abledata.com	Provides information on assistive technology and rehabilitation equipment available from domestic and international sources to consumers, organizations, professionals, and caregivers within the United States.
National Council on Independent Living (NCIL) *Not just responding to change, but leading it*	1916 Wilson Boulevard, Suite 209 Arlington, VA 22201 Voice: 703-525-3406 TTY: 703-525-4153 e-mail: ncil@ncil.org Web: www.ncil.org	There are several hundred Centers for Independent Living (CIL) across the country. CILs are community-based resource and advocacy centers managed by and for people with disabilities, promoting independent living and equal access for all persons with physical, mental, cognitive, and sensory disabilities.
The Arc of the United States *Supports issues of cognitive disabilities*	1010 Wayne Ave., Suite 650 Silver Spring, MD 20910 Voice: 301-565-3842 e-mail: info@thearc.org Web: www.thearc.org	The Arc of the United States works to include all children and adults with cognitive, intellectual, and developmental disabilities in every community.
Office of Disability Employment Policy (ODEP) *Directory for (50) States' assistance*	1331 F Street, NW, Suite 300 Washington D.C. 20004 Voice: 202-376-6200 TTY: 202-376-6205 Web: www.dol.gov/dol/odep	State liaisons are the contacts on disability issues in each of the respective states. For the resources provided by the liaison in your state, go to www.dol.gov/dol/odep/public/directory, and then *state abbreviation*. Click on the abbrevation for your state.
JAN, The Job Accommodation Network *Comprehensive resource*	PO Box 6080 West Virginia University Morgantown, WV 26506-6080 Voice & TTY: 1-800-526-7234 e-mail: jan@jan.icdi.wva.edu Web: www.jan	Provides information and consulting on accommodating people with disabilities in the workplace. JAN Accommodation Toolbox offers free publications. For example, evacuation procedures, effective communication for individuals who are deaf or hard of hearing, ergonomics, and disability etiquette.
Eastern Paralyzed Veterans Association (EPDA) *Technical assistance on accessibility and building codes*	111 West Huron Street Buffalo, New York 14202 Voice: 800-795-3619 e-mail: bblack@epva.org Web: www.epva.org	Provides information on accessibility and life safety for persons who use wheelchairs, with an emphasis on the requirement of model building codes and standards. Provides free publications and a monthly newsletter on issues affecting paralyzed veterans.

Appendix C

Disaster Mitigation for Persons with Disabilities

For the 54 million Americans with disabilities, and millions of others around the world, surviving a disaster can be the beginning of a greater struggle. Whether an individual with a disability requires electricity to power a respirator, life-sustaining medication, mobility assistance, or post-disaster recovery services, relief organizations and rescue personnel must be prepared to address the needs of that individual in the hours and days following a disaster.

Similarly, efforts to accommodate disabled Americans frequently ignore disaster preparedness and response. As a result, too few disaster response officials are trained to deal effectively with people with disabilities, and too few disabled Americans have the knowledge that could help them save their own lives.

Seven key principles should guide disaster relief:

1. Accessible Disaster Facilities and Services

Communications technology is vital for people with disabilities during a disaster to help assess damage, collect information, and deploy supplies. Access to appropriate facilities—housing, beds, toilets, and other necessities—must be monitored and made available to individuals with disabilities before, during, and after a disaster. This access also must be ensured for those who incur a disability as a result of a disaster. Appropriate planning and management of information related to architectural accessibility improves the provision of disaster services for persons with disabilities.

2. Accessible Communications and Assistance

As communications technology and policy become more integral to disaster relief and mitigation, providing accessibility to the technology for people with disabilities becomes more essential. For example, people with hearing impairments require interpreters, TDD communications, and signaling devices. In addition, written materials must be produced on cassette tape, on CD-ROM, or in large print for people with visual impairments. People with cognitive impairments, such as those with developmental disabilities, Alzheimer's disease, or brain injury, require assistance to cope with new surroundings and to minimize confusion factors. It is crucial that people with disabilities help develop accessible communications and reliable assistance technologies.

3. Accessible and Reliable Rescue Communications

Accessible and reliable communications technology is critical to ensuring fast, effective, and competent field treatment of people with disabilities. Current satellite and cellular technology as well as personal communication networks permit communication in areas with a damaged or destroyed communication infrastructure. Communications technologies can assist field personnel in rescue coordination and tracking and can be combined with databases that house information on optimal treatment for particular disabilities or that track the allocation of post-disaster resources.

4. Partnerships with the Disability Community

Disability organizations must join with relief and rescue organizations and the media to educate and inform their constituents of disaster contingency and self-help plans. A nationwide awareness effort should be devised and implemented to inform people with disabilities about necessary precautions for imminent disaster. In the event of a sudden natural disaster, such a program would minimize injury and facilitate rescue efforts. In addition, more young people with disabilities should be encouraged to study technology, medicine, science, and engineering as a way of gaining power over future technological advances in disaster relief and mitigation.

5. Disaster Preparation, Education, and Training

Communications technologies are crucial for educating the public about disaster preparedness and warning the people most likely to be affected. Relief and rescue operations must have the appropriate medical equipment, supplies, and training to address the immediate needs of people with disabilities. Affected individuals may require bladder bags, insulin pumps, walkers, or wheelchairs. Relief personnel must be equipped and trained in the use of such equipment. In addition, relief personnel should provide training, particularly for personnel and volunteers in the field, on how to support the independence and dignity of persons with disabilities in the aftermath of a disaster.

6. Partnerships with the Media

Many natural disasters can be predicted in advance. Disaster preparedness for people with disabilities is critical in minimizing the impact of a disaster. The media—in partnership with disability and governmental organizations—should incorporate advisories into emergency broadcasts in formats accessible to people with disabilities. Such advisories alert the public, provide a mechanism for informing rescue personnel of individual medical conditions and impairments, and identify accessible emergency shelters. The creation and repetition of accessible media messages is critical for empowering people with disabilities to protect themselves from disasters.

7. Universal Design and Implementation Strategies

Designing universal access into disaster relief plans, far from being a costly proposition, can pay off handsomely. As accessible communications tools become more widely available, their price will decrease. In addition, a universal design approach to meeting the needs of people with disabilities before and after a disaster will benefit many people without disabilities, such as the very young or the aged. A look at existing agreements among relief organizations and local, state, federal, and international governments will offer guidance in developing effective strategies for universal design and implementation plans. The federal government's role has yet to be defined, but it could encourage or even mandate universal design and set standards. For example, the federal government could provide guidelines for evacuation plans or pre-disaster warning periods.

From a report by The Annenberg Washington Program written in collaboration with the President's Committee on the Employment of People with Disabilities by Annenberg Senior Fellow.

Appendix D
TTY – Commonly Used Abbreviations

It is common for TTY callers to use many abbreviations in their TTY conversations. This is used to minimize keystrokes and long distance charges on the TTY. These abbreviations are also used to supplement modern abbreviations, postal codes, and "emotions" such as ":-)" on 2-way wireless text messaging as well as Internet instant messaging applications. (★) are the basics in all TTY conversations and italics are more applicable to weather or emergency situations. Be sure to find out any abbreviations common to your locale, which may not be known nationwide.

Resource for This List

TDI - Telecommunications for the Deaf, Inc.
8630 Fenton Street, Suite 604, Silver Spring, MD 20910-3803
301-589-3006 (TTY); 301-589-3797 (fax); 301-589-3786 (v)
Web: www.tdi-online.org/quicklists1.html

A
ABT About
ADA Americans with Disabilities Act
AM Morning
AMBL, AMBU *Ambulance*
ANS Answer
ANS MACH Answering Machine
ASAP As soon as possible

B
BEC, CUZ Because

C
CA Communication Asst. (Relay Operator)
CC (Closed) Captioning
CD, CLD Could
CI Cochlear Implant

CLR Clear
CU See you!
CN Can

D
DOB *Date of Birth*
DOC, DR *Doctor*

E, F, G
★GA Go Ahead

★GA SK, GA to SK Anything else?
GOVT Government

H
HAND Have a nice day!
★HD, HLD Hold
HOSP *Hospital*

I
IMPT Important
INFO Information

INT, TERP Interpreter
ILY I love you

J, K
KIT Keep in touch!

L
LD Long-distance
LTR Letter

LV Leave

M
MIN Minute
MSG, MSSG Message
MTG Meeting
MYOB Mind your own business!

N
N and **NBR** Number
NITE Night
NP, NO PBLM No problem!

O
OK Okay
OIC Oh, I see
OPR Operator
OXOX Hugs and kisses

P
PH Phone
PLS Please

Q
★Q, QQ, QM Question mark

Appendix E

References

Photographs and Drawings:

The following organizations have generously provided permission to reprint photos and drawings for publication:

- Emergency Education Network (EENET) Federal Emergency Management Agency (FEMA) video: *Meeting the Special Needs of the Disabled in Evacuation and Sheltering Systems*; March 22, 1989.

- National Research Council of Canada, *Evacuation Techniques for Disabled Person*; March 1983

- Independence Dogs, Inc. (IDI)
  Service Dogs for the Mobility Impaired
  146 State Line Rd.
  Chadds Ford, PA 19317
  Voice: 610-358-2723
  Web: www.independencedogs.org

- Garaventa Canada, Ltd.
  Web: www.evacutrac.com/index.html

- EVAC+CHAIR® Corporation
  PO Box 2396
  Voice: 212-369-4094
  e-mail: sales@evac-chair.com
  Web: www.evac-chair.com/

- Servison, Inc.
  982 N.E. 126th Street
  North Miami, FL 33161
  Voice: 305-899-1996

- Baldwin Emergency Medical Service, Inc.
  One Readshaw Way
  Pittsburgh PA 15236
  Voice: 412-884-0666
  Web: www.baldwinems.com/

- Western PA School for the Deaf
  300 East Swissvale Avenue
  Pittsburgh, PA 15218
  Voice: 412-371-7000
  Web: www.wpsd.org/home.htm

Equipment and Product Information:

- Telecommunications for the Deaf, Inc. (TDI)
  8630 Fenton Street, Suite 604
  Silver Spring, MD 20910-3803
  Voice: 301-589-3006
  TTY: 301-589-3797
  Web: www.tdi-online.org/

- Delta Society®
  580 Naches Avenue, SW, Suite 101
  Renton, WA 98055-2297
  Voice: 425-226-7357
  Web: www.deltasociety.org

- EVAC+CHAIR® Corporation
  PO Box 2396
  Voice: 212-369-4094
  e-mail: sales@evac-chair.com
  Web: www.evac-chair.com/

- Contra Costa County Fire Protection District
  Station 6, Battalion 2
  Pleasant Hill, California

- National Organization on Disability (N.O.D.)
  910 Sixteenth Street, N.W., Suite 600
  Washington, DC, 20006
  e-mail: ability@nod.org
  Web: www.nod.org

www.ingramcontent.com/pod-product-compliance
Lightning Source LLC
Chambersburg PA
CBHW081410170526
45166CB00010B/3283